BACKYARD BEEKEEPING

WE TAKE THE STING OUT OF BEEKEEPING

Jason & Mindy Waite
Mike & Rhonda Wells

HARVEST LANE HONEY
H·L·H·

© 2016 by Sourced Media Books, LLC.

Sourced Media Books, LLC.
15 Via Picato
San Clemente, CA 92673

ISBN 978-1-937458-83-6
www.sourcedmediabooks.com

CONTENTS

FOREWORD
A Hobby with Purpose

One of the most rewarding aspects of backyard beeping is that it's a way for each of us to do our small part to reverse the alarming and unexplainable decline of North American bee populations since the mid-2000s, a phenomenon known as colony collapse disorder (CCD). More than 10 million beehives were inexplicably wiped out between 2007 and 2013 alone.

Tending to backyard beehives also is one of the most therapeutic, peaceful hobbies that we can adopt. In fact, as long as you're suited up properly and handle yourself properly, your bees probably won't even know you're there. And contrary to popular belief, your bees probably aren't going to sting you or swarm you.

Finally, with proper care and some patience, you can harvest bountiful yields of your own honey and beeswax to share with family and loved ones. We have found that the absolute best part about backyard beekeeping is that the sweet reward of being able to share the joyful, all-natural taste of Mother Nature's candy.

Over the years, we have turned our passion for beekeeping into a family business that we named Harvest Lane Honey. We've focused on developing a high-quality alternative to the classic—and classically flimsy—Langstroth beehive design. With Jason's architecture and engineering background, the design of our Trademark wooden hive components can withstand punishing rains, crushing snowfalls, and moisture and rot. Today, the innovative Langstroth design we developed is available at major retailers nationwide.

Beekeeping truly is a hobby with purpose, and all of us at Harvest Lane Honey look forward to sharing all the joys of this sweet harvest with you. Happy beekeeping!

Jason & Mindy Waite
Mike & Rhonda Wells

A BRIEF HISTORY OF *Beekeeping*

Honey harvesting is almost as old as human history, dating back thousands of years. In fact, the tradition of using smoke to rob wild bees of their honey is depicted in ancient cave drawings.

It is quite likely that our ancestors' first encounters with wild beehives in Africa were punctuated by painful stings. Once those early humans discovered how to control fire, however, they learned how to exploit the bees' natural response to the threat of fires. Smoke is a signal to bees that they should rush to fill their stomachs with honey and then abandon the hive. In this rush to vacate, bees also tend to be less interested in protecting the colony with their stingers.

Over time, hunting for honey evolved into beekeeping, with ancient civilizations building hives to house colonies collected from the wild. Archaeologists, for example, found 30 fully intact beehives dating back to the early ninth century BCE in the ruins of Rehov, Israel, an ancient town in the Jordan Valley. These manmade beehives, discovered in 2007, were constructed of straw and unbaked clay, and arranged in orderly rows, showing that our interest in beekeeping is well over 3,000 years old!

Beekeepers have learned tremendous amounts of information about honey production and the value of beeswax through the centuries. Beeswax has been used in candles, jewelry making, encaustic painting, and waterproofing, among other functions. Our beekeeping predecessors also learned about bees' important contribution to more fruitful and flavorful harvests. For centuries, farmers have kept bees in hollow logs, pots, and wooden boxes alongside their fields and orchards to promote pollination of their crops—and, of course, produce honey for their families.

One of the most significant advances in beekeeping came in the 19th century, when researcher Francois Huber discovered that the optimal space in which bees like to work is one-quarter to three-eighths of an inch wide. Armed with knowledge of this so-called "bee space," the Rev. Lorenzo Lorraine Langstroth designed a revolutionary beehive that was built to ensure maximum comfort for bees as they built and sealed their honeycombs.

Langstroth designed a hive with removable frames based on those measurements. By devising a way to encourage bees to build their honeycombs on a frame that could be easily removed, then returned to the hive after the honey was extracted, he found a way for people to co-exist with bees and keep their colony alive while still harvesting valuable honey and beeswax. He published his patented beehive design in 1853, a design that is still used in many parts of the world to this day.

75%
of the world's beekeepers use Langstroth hives.

At least 75% of the world's beekeepers are currently using Langstroth hives, and that number is even higher in North America for a number of reasons.

The highly efficient Langstroth hive quickly contributed to the bee boom of the 20th century. Beekeeping became a profitable business as well as an accessible hobby. The symbiotic relationships between keepers and their bees led to conscientious gardening, specially designed areas for hives, industrial beekeeping operations, and hybridization of domesticated bees. Colonies began to live longer because they were more resistant to disease, weather, and other natural threats. Bee populations soared and honey became more available than at any other time in human history. Farmers grew to rely on commercial beekeepers and their traveling hives to pollinate crops.

Today, an estimated one-third of our food supply depends on insect pollination, most of which is accomplished by honeybees. For this reason, most professional beekeepers are now engaged in the business of contract pollination; honey production has become a secondary goal.

Because of bees' economic importance, beekeepers and farmers are now focused on how to protect their hives from modern dangers, including Colony Collapse Disorder (CCD), Israeli Acute Paralysis Virus (IAPV), and other viruses and infections that threaten the world's bee population.

In recent years, we have seen precipitous declines in worldwide bee populations. In 2014, North American bees experienced their largest die-off in recorded history. In healthy circumstances, honeybee colonies should thrive year after year. What is especially unusual about these losses is that many occurred in mid-summer rather than during harsh winter conditions. Researchers have postulated that the high die-off rates may be due to increasing loss of pollinating species and the plants they rely on for their survival. But those are only guesses, and the reasons remain unclear.

As society works to better protect bees by banning certain insecticides in some markets and developing better ways to protect pollinating bee species from disease, we can all do our own small part through backyard beekeeping. Backyard beekeeping promotes the well-being of native and domesticated bee species, and more broadly, supports the health of our environment. Happily, backyard beekeeping is easier than ever. In the next few chapters of this book, we'll teach you everything you need to know to be successful at it.

THE WIDE WORLD OF
Bees

Most people think that a bee is a bee. But there are actually about 4,000 bee species in the United States alone, and the insect family that bees are a part of encompasses as many as 25,000 species worldwide. Indeed, the average American backyard is a blooming feast for dozens of species of wild bees. Hence, when we encourage local wild bee species, we are promoting the beauty and success of not only our own gardens, but also the endangered species that play an important role in local ecology.

4,000

Bee species in the United States alone, and the insect family that bees are a part of encompasses as many as 25,000 species worldwide

SOLITARY WILD BEES

When most people think of bees, they probably think of the impressive bumblebees or carpenter bees that drift through their flowerbeds, or possibly the honeybees from nearby hives. Many, however, fail to notice solitary bees like the mason or leafcutter bee. Solitary bees are friendly bees that are not a threat to people, pets, or other pollinating species. They are nonaggressive and hyper-efficient pollinators that contribute to the success of plants and trees that bloom in early spring, and greatly enrich fruit harvests. Eighty percent of our food crops depend on bees for successful harvests.

Solitary wild bees have a fascinating life cycle. While bumblebees often nest underground in abandoned mouse nests or compost heaps, many species of solitary bees live in crevices, old wood, or the hollowed-out stems of plants. Active for just a few weeks during the blooming season, solitary bees have much shorter lifespans than honeybees. During that time, they will pollinate thousands of flowers and blossoms while building nests for next year's generation.

Solitary bees build cavity-nesting burrows that contain several cells, each containing a lump of pollen and an egg. The larva usually takes one year to develop into an adult bee. Mason bees search for hollow stems or reeds to enclose eggs behind a tiny mud wall. Leafcutter bees follow a similar process, encasing their young behind walls carefully cut from fibrous leaves.

If you want to attract native solitary bees, all you need to do is fulfill their two basic needs: food and shelter. First, you need flowers to provide food for bees (in the form of nectar and pollen). Ideally, your garden should be filled with many different kinds of trees or flowers that bloom at different times of the year, so there is always a source of nectar and pollen available. As you tend to your garden, the single most important thing you can do to keep your solitary bees coming is to use pesticides responsibly. If you must spray

around the gardens and trees they visit, do so at dawn or dusk when bees are safely in their hives. Follow all package instructions, and never spray flowering plants with a pesticide that is identified as dangerous to pollinating species.

Second, you need to provide safe nesting spaces to attract solitary wild bees. Just as you might put out nest boxes for birds, you can set up nest boxes for wild bees. By hanging a wild bee house, you are replacing the naturally occurring nesting opportunities that we tend to clear away from our gardens and yards in the fall.

In most cases, bee nest boxes are wooden enclosures filled with hollow reeds. Harvest Lane Honey offers a sturdy bee house that affixes to a wall or fence and provides sufficient protection from wind and weather, as well as natural 8mm reeds that wild bees prefer. You will need to replace these reeds every couple of years to protect wild bees from fungus, infection, and parasites.

You'll want to hang your solitary bee house in a secure location that receives morning sun and is protected from sprinklers and

rain. Ensure that there is a constant mud or leaf source nearby, depending upon the bee you are hosting. To protect your bee eggs as they develop, you can remove the egg-filled reeds from the bee house in the fall and store them in a cool, dry place indoors (like a refrigerator) until spring, ensuring they are spared from hungry birds that raid insect nests when food is scarce during the winter. In early spring, as temperatures approach the bees' comfort zone, simply return the reeds to the cleaned bee house, and the cycle will begin anew.

If you want to attract native solitary bees, all you need to do is fulfill their three basic needs: **FOOD, SHELTER, & WATER**

If solitary bees aren't finding their way to your bee house, an alternative option is to order live solitary bees. Your bees will arrive in nesting tubes, ready to hatch when the temperature is right. And they will hatch! If your bee house isn't ready when they arrive, keep these tubes in a closed, ventilated container in your refrigerator until you can safely place them outdoors.

If you buy solitary bees, you should carefully research the best species for your region, as well as their active season. Bees that awake in early spring are going to be searching for blooming flowers and trees, so if you release them in the heat of summer, they won't thrive. It's much better to get your bees early than too late in the season. Because they naturally remain dormant during cold winters, you can easily keep the nests in a cool, dry, protected place, such as a closed container in a refrigerator, a basement cold storage room, or an undisturbed location in a garage.

HONEYBEES

The European honeybee, also known as the common or western honeybee (Apis mellifera), is probably the most recognizable insect in the world, earning its name from its unique ability to produce large amounts of honey. It is believed that the honeybee originated in Africa and spread to northern Europe, India, and China. It is not native to North America, but hives were brought over by ship to provide honey for the first colonists. The honeybee has since thrived, and is now distributed worldwide.

The life span of an individual honeybee is relatively predictable, depending on its role within the colony. Honeybees live in sophisticated, well-organized societies comprised of between 50,000 and 80,000 individual bees, each performing different roles to ensure the success of the hive. The population of a typical honeybee colony might break down as follows:

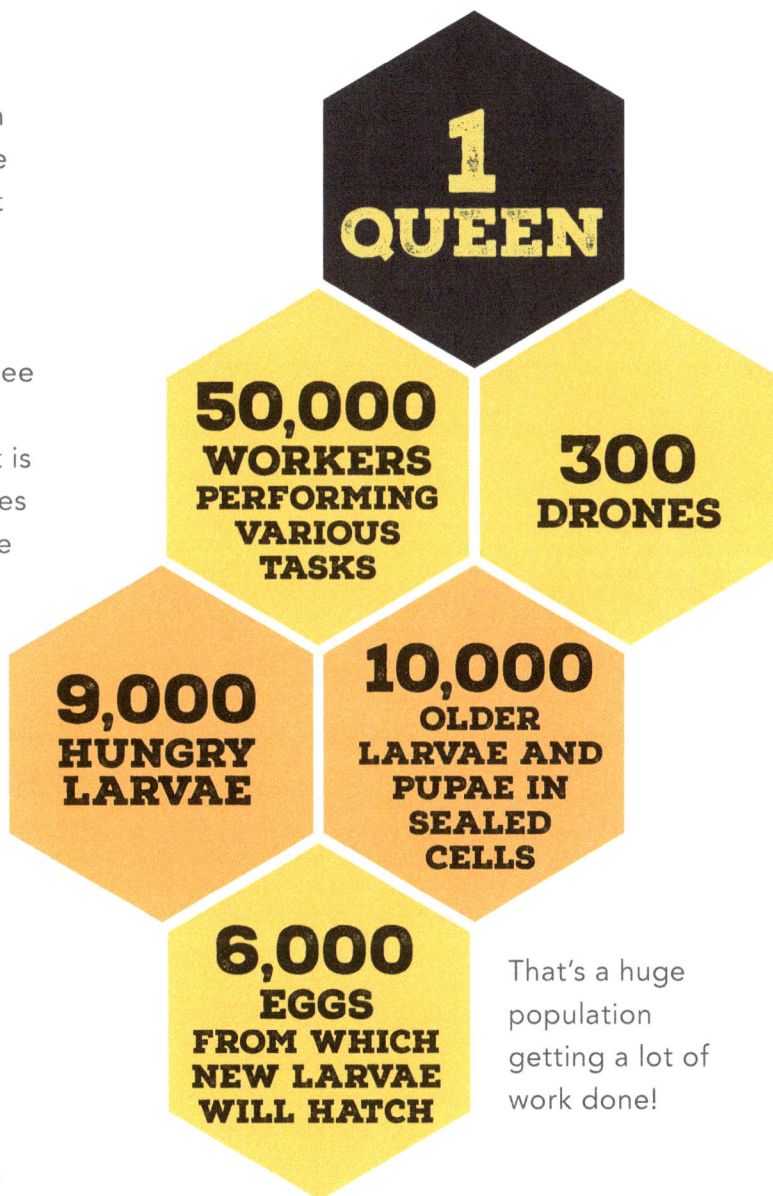

1 QUEEN

50,000 WORKERS PERFORMING VARIOUS TASKS

300 DRONES

9,000 HUNGRY LARVAE

10,000 OLDER LARVAE AND PUPAE IN SEALED CELLS

6,000 EGGS FROM WHICH NEW LARVAE WILL HATCH

That's a huge population getting a lot of work done!

Queen

Drone

Worker

The queen bee is identified by her noticeably elongated body and straight stinger with no barbs. Drones are identified by their stout bodies and large eyes. Neither the queen nor drone will die if they sting. Worker bees have a barbed stinger and hind legs specialized for collecting pollen. Each worker bee's legs are flattened and covered with long fringed hairs that form a pollen basket.

While every member of the colony is essential, the worker bees are responsible for the day-to-day operations of the hive.

Once a worker bee emerges from the pupae state, she immediately gets to work, removing waste from nearby cells. She produces a naturally disinfecting material that makes the cells ready for new eggs. After a few days, she becomes a nurser bee, feeding the larvae with pollen and honey. Later, she will feed royal jelly to the queen. At approximately 16 days of age, she will begin secreting wax from her abdomen for building honeycomb. Several days later, she will perform guard duties, defending the entrance of the hive or nest from predators.

Later, the worker bees leave the hive to begin foraging for the nectar and pollen that feed the colony. Additional duties of these diligent workers may include fanning their wings to regulate interior temperature, removing dead bees and larvae, and carrying water into the hive.

The time of year in which a worker bee is born will also define her life span and her role in the colony. Workers born in spring/summer have shorter, busier lives, while those born in autumn may live longer, but must survive harsher winter conditions. Winter-born bees must care for larva and huddle together around the queen in order to keep warm, then emerge ready to begin foraging in the early spring.

The drones make up the small male population of the colony. These larger bees live up to four months and have a single task: to mate with the new queen or a virgin queen. Immediately after mating, they die.

The primary responsibility of the honeybee queen is to lay eggs—in fact, thousands of them over the course of her life. A newly hatched virgin queen will mate early in her life during a single mating flight with 6–10 male bees, or drones. After this flight, the queen stores up to 100 million sperm within her oviducts. If a queen runs out of sperm in her lifetime, a new queen will be raised to mate and produce her own colony.

When a colony becomes too big or the hive unsuitable, the queen establishes a new colony by swarming. Swarming is typically a perfectly healthy and normal behavior. When the time is right, a new queen is produced, and the original queen takes off with half of the hive's population. While this mass of swarming bees might look frightening, they're actually tremendously docile and are ideally suited for placement in another hive. They will naturally establish a new colony in a new location, leaving the new queen behind with the existing colony.

Honeybee queens determine whether a particular egg gets fertilized. Unfertilized eggs become drone honeybees, while fertilized eggs develop into female workers and queens. Female workers do not mate, but they can lay infertile eggs, which can become drones.

As a beekeeper, you can visually identify the roles of the larva growing inside your hive. Worker bees hatch from flat cells arranged in an organized pattern near the center of the frame. Drones grow inside larger cells. A queen grows inside a much larger cell (peanut shaped), the location of which will indicate your colony's intention. If a supersedure is underway and a queen is being replaced, the queen cell will be found in the center of the frame. This queen will hatch to replace an aging or unwell queen in the hive. If a queen cell is growing near the top or the bottom of your frame, your colony is planning to swarm.

If all is well, a healthy honeybee queen can live as long as four years. This is much longer than wild bumblebee queens or solitary bee species. A productive queen will lay an unimaginably large number of eggs during her lifetime: about 2,000 eggs per day (in the summer time). If she lays too few eggs or is hurt, her workers will replace her with a new queen, called supersedure.

By understanding the members of your bee colony and their responsibilities, you will be better able to care for them from one year to the next. The good news is that they typically don't need much more than a clean, comfortable home close to reliable food and water sources.

Of course, the devil is in the details, and that's what we'll explore over the next few chapters.

BEFORE YOU GET STARTED

Deciding to take up beekeeping is like embarking on a journey to a completely different culture. To keep your bees happy and safe, you'll need to understand what your bees need from you, as well as what tools and tasks you need to master to make beekeeping safer and more productive for you and your bees. Your goal should be to learn a little more about your bees' complex community every day and to develop a symbiotic relationship with your hive.

> *To keep your bees happy and safe, you'll need to understand what your bees need from you.*

Before you buy anything, you should do research on local laws regarding keeping bees. Your county or municipality may have restrictions, such as how many hives you may keep and which varieties of bees are acceptable. There may even be ordinances prohibiting beekeeping altogether. In many states, there are regulations that require beekeepers to register their apiary locations and pay a small annual registration fee. Many of these policies are intended to protect the greater bee population, serving as a means for controlling diseases that can destroy colonies. This information is easy to find online, and knowing it in advance can save you considerable time and money.

Next, you want to explore your neighborhood and its potential for hosting more than just bees on your own property. For example, you may find that you have neighbors happy to house your hives near their fruit trees or in a large, unused back lot. Or perhaps your apartment manager

will be delighted to include your hive on a rooftop garden. If you stumble upon a community garden, you might be surprised at how grateful these gardeners would be to gain an extra army of pollinators.

You also should tap into the community of beekeeping clubs and/or organizations in your area. Experience is the best teacher, and connecting with your fellow beekeepers will provide answers and assistance when you need them most. They will know how to address challenges specific to your climate, they'll be aware of local threats, and they'll share your enthusiasm for this fascinating species. They can also provide recommendations for community education events, voluntary efforts, and policy change opportunities that can promote beekeeping in your region, contribute to the success of your hive, and better protect the entire pollinating insect population. In the small but vibrant world of beekeeping, your contribution is very important. Get involved!

CHOOSING A HIVE

Choosing a hive can seem intimidating, but this step is actually much simpler than most beginners realize. When researching pre-built hives, you should be looking for high-quality hives able to withstand considerable weight and extreme weather. Look for hive boxes and frames constructed from premium ¾" wood frames and protected with two coats of exterior paint. Although a pitched metal top is not necessary, it will improve the appearance and durability of your hive and save you money in the long run. A sliding screened bottom board will protect your colony from cold air in the winter.

The hive itself is a complex piece of architecture. While there are a number of different hive designs, we are strong proponents of the classic Langstroth hive. This the most widely used hive in the United States, and it is gaining popularity worldwide.

Harvest Lane Honey is pleased to offer all-in-one backyard beekeeping kits that save you the trouble of building your own hive and researching the necessary equipment. Our kits are a terrific entry point for first-time beekeepers.

CHOOSING A LOCATION

Choosing the right location for your hive will contribute to the success of your colony, as well as to your enjoyment of beekeeping. There are many factors to consider, chief among them your own ease of access to your hive. Here are some important tips for choosing a location for your hive:

- If you are using a screened bottom board for ventilation, you will want a location that is free of obstructions on the back side of the hive. Approaching the hive from behind its entrance is safer for you and less startling to the bees guarding its entrance.

- If you live in an urban or suburban area and you feel there is potential danger to either your bees or your neighbors, you should aim to place the hive in a relatively concealed area, preferably one that is fenced off.

- Ideally, you should place your hive in a location that is protected from harsh winds and exposed to morning sun, as the sun will warm the hive and get it active earlier in the day.

- Your hive should be placed on level ground. If the hive stands at a slight slant toward its front entrance, water will run out of the hive. You can place garden tiles, rocks or scrap wood under your hive stand to create the optimal angle.

- You should position your hive to be near a food source (i.e., flowering plants and trees).

- Because bees need water and will frequent the closest reliable source, you want to ensure you have a water source in close proximity. You can create one easily by placing a shallow-sloped water source near the hive, such as a pie tin weighted down with small pebbles. A water source you can control is crucial, because if the closest water source is your neighbor's pool or bird bath, your bees will be more than willing to pay it a visit, regardless of how inconvenient it is for your neighbors.

CHOOSING A HONEYBEE VARIETY

If you're just getting started as a beekeeper, you will want to choose a honeybee variety that is well-suited for your climate, skill level, and desired honey production. Honeybee types vary only slightly in appearance, but greatly in temperament and hardiness. There is no "best" strain of bee, as the traits favored by one beekeeper may differ significantly from another's choice. If you're not sure which variety you want, you should consider the following four popular varieties:

ITALIAN HONEYBEES (APIS MELLIFERA LIGUSTICA):

This is the most commonly kept variety of honeybee and a great choice for beginners because they tend not to swarm and have gentle dispositions. This variety also does very well in hot climates, making it a good choice for beekeepers in the southern United States. Italian honeybees have a habit of building large colonies that produce tremendous honey stores, which sometimes mean they require feeding during the winter.

CAUCASIAN HONEYBEES (APIS MELLIFERA CAUCASIA):

This is a hardy bee, able to forage in colder temperatures that would not suit other varieties. Caucasian honeybees are slow to start in the spring, but you can count on a strong population. While generally calm, they are difficult to subdue if upset.

CARNOLIAN HONEYBEES (APIS MELLIFERA CARNICA):

Keep a close eye on these extremely gentle bees. They may not require winter feeding, but they have a propensity for swarming and splitting off into new hives. They are, however, hardy in colder climates. When happy and comfortable, they produce great honeycomb, even early in the spring.

RUSSIAN HONEYBEES (APIS MELLIFERA):

This bee is more expensive, but also more resistant to mites and challenging climates like cold winters than other varieties are. Because they are more aggressive than other varieties, they are probably not an ideal starter bee, but definitely a good choice for more established beekeepers who live in harsh climates.

To order bees, contact a local beekeeper or a retail bee supplier. Bees are shipped in what is called a package, which usually consists of about 10,000 bees (sometimes labeled as three pounds of bees) and a mated queen in a cage.

Essential Tools OF THE TRADE

The costs associated with becoming a first-time beekeeper are unfortunately not insignificant. In order to keep your bees alive and safe, you'll need to invest in a number of tools that have stood the test of time, including frame grips, a hive tool, a good smoker, and additional honey supers. Top-quality tools recommended by experienced beekeepers are essential to making your beekeeping experience safe and seamless.

PROTECTIVE CLOTHING: To keep yourself safe while interacting with your bees, you'll need to invest in a protective bee suit with elastic closures at the ankles and wrists. This suit should include a veiled hood to keep bees away from your face. We also recommend sturdy goatskin gloves designed specifically for beekeeping, and white rubber boots for further protection. It is essential that you wear a veil at all times, as this netted headgear protects your face when working with the hive. Buy the best-quality clothing you can find to spare yourself the painful experience of learning just how many layers a defensive bee's stinger can penetrate.

Jacket

Gloves

Bee Suit

SMOKER: Nothing calms an agitated hive of bees like a few puffs of cool white smoke. The smoke activates the bees' instinct to consume large amounts of honey when they sense the threat of fire. While a smoker is a powerful tool, it takes some practice to learn how to use properly. Practice lighting and maintaining your smoker before you have your hive open. Light a small amount of fuel at the bottom, then use the bellows to stoke the embers while you pack the canister full. To prevent your smoker going out at exactly the wrong time, always use high-quality burlap or smoker pellets as fuel.

HIVE TOOL: This is a specialized pry-bar with a sharp edge. It's useful for many beekeeping tasks, but is mainly used for separating and removing frames from a hive body. Inspections would be possible without this important tool.

POLLEN PATTIES: Supplemental feeding of bees in the spring and fall is a common practice, and you'll need a feeder to ensure your bees have access to this food. More important than the design and placement of the feeder is what you put in it. At Harvest Lane Honey, we have developed a highly nutritious line of foods for supplemental feeding. Our products also contain directions for properly supplementing these foods with nutrients and medication when needed.

UNCAPPING/ELECTRIC KNIFE: This tool, used to remove the wax covering that bees deposit over their honey, is an absolutely essential implement for extracting honey. A standard uncapping knife is heated in hot water, but you also can buy an electrically heated uncapping knife—it's a nice luxury during your honey harvest.

SCRATCHER: This sharp, comb-like tool finishes the job of opening honeycomb by puncturing shallower cells, which allows honey to flow freely.

HONEY EXTRACTOR AND CONTAINERS: The honey extractor is a centrifuge that spins uncapped frames of honey at a high speed until all of the honey clings to the sides of the chamber. (Extractors are explored in more depth later in this book.) The honey then drips to the bottom spout for easy pouring through cheesecloth or a metal sieve into food-grade buckets or jars. If you've taken good care of your bees, you'll need several containers on hand to capture all of your sweet harvest.

COMPONENTS OF A
Langstroth Hive

The classic Langstroth hive is an optimal choice for most beginning beekeepers. The screened bottom board and chimney effect are designed for superior airflow, temperature management, and pest control, optimizing the chances of raising a healthy, thriving colony of honey-producing pollinators. The Langstroth hive is also compatible with a plethora of beekeeping and honey extraction equipment.

If you're looking to buy a Langstroth hive, be very conscientious about looking at the quality of construction. A well-built Langstroth hive will last you a lifetime, so this is no place to scrimp on quality. You should look for well-built components that can stand up to the rigors of your climate: full wood construction, metal covers with drip edges, and two coats of exterior paint.

At Harvest Lane Honey, we've improved on the original Langstroth design, creating a hive that can hold up to the harshest climates in North America. In this chapter, we will provide an overview of all of the components that make up a Langstroth hive.

ELEVATED HIVE STAND: A sturdy hive stand is used to elevate the hive off the ground, which improves air circulation and decreases risk of mold or flooding from damp soil. In addition, the stand raises the hive above grass and other vegetation that can slow your bees' ability to get in and out of their hive. Finally, having a stand means you won't have to bend over as much as you're tending to your hives.

BOTTOM BOARD: This is the floor of the beehive. It consists of several rails that serve as a frame around a solid piece of wood, protecting the colony from damp ground. More and more beekeepers are using screened bottom boards to improve ventilation and to help control and monitor varroa mites.

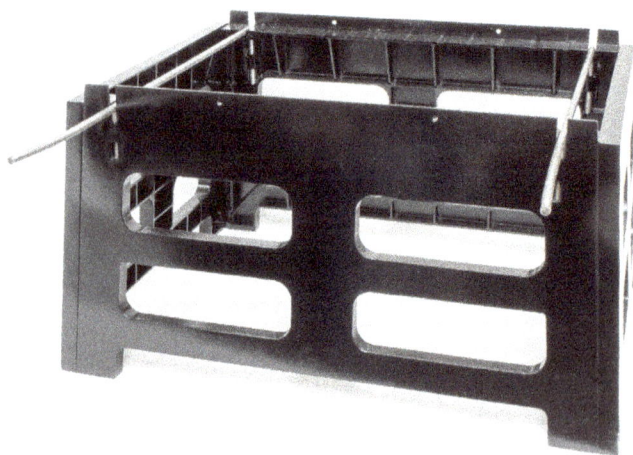

ENTRANCE REDUCER: An entrance reducer gives you the ability to limit bee traffic in and out of the hive. You can put it in place and remove it as needed. For example, you can use it to control ventilation and temperature during cooler months, and as your colony grows, you can use this simple device to help your guard bees better protect their hive. Our reducer is placed in front of the brood box.

DEEP HIVE SUPER: This is a sturdy wooden box that contains frames of honeycomb. When setting up a Langstroth hive, you typically would stack two deep hive bodies on top of one other, like a two-story condo (start out using one deep hive and when it has filled 8 of the 9 frames, then add your next deep hive). The bees use the lower deep as the nursery or brood chamber, and the upper deep super as their pantry or food chamber, where they store most of the honey and pollen they will eat over the winter. Frames in the upper box will have brood in the center and essential honey stores around the outside edges.

QUEEN EXCLUDER: A must for beginning beekeepers, the queen excluder allows you to find the queen easily during hive inspections. The queen excluder should be placed above the deep brood and honey supers, which will prevent the queen from laying eggs in honey that will be harvested.

HONEY SUPER: Honey supers are used to collect surplus honey. As the bees collect more honey, you can add more honey supers to the hive, stacking them on top of each other like stories on a skyscraper. Inside your shallower boxes is the honey that you can safely harvest from your bees. (The honey in the deep hive body must be left for the bees.) Honey supers are identical in design to the deep hive bodies, but the depth of the supers is shallower, making the frames much easier for you to handle when harvesting honey. Keep in mind, however, a medium super weighs around 55 pounds when packed full.

FRAMES: Frames are wooden and come with a single sheet of beeswax foundation. The bees build their honeycomb onto this foundation. Frames typically come in two basic sizes: deep and medium, corresponding to deep hive bodies and medium honey supers. Frames are removable from the hive, allowing you to easily inspect, manipulate, and manage your colony.

INNER COVER: The inner cover is a shallow, telescoping tray with a ventilation hole in the center; because it is telescoping, bees are less likely to cement it into place with propolis, making it easier for you to maintain your hive.

OUTER COVER (FLAT TOP): The outer hive cover is a weatherproof material used to protect bees from the elements. Like the roof on your house, the outer hive cover ensures that your bees' home is waterproof. A quality cover, like Harvest Lane Honey's product, is essential to maximizing the life of the hive. To protect the hive from moisture runoff, you should look for a cover with a drip edge.

Transferring Bees
TO THEIR NEW HOME

The moment that your queen and colony arrive is one of the most satisfying experiences of beekeeping: You get to meet your bees for the first time.

Your bees will arrive in a small wooden box with screened sides and a can filled with bee feed syrup. Although bees historically were shipped directly to individuals by mail, now it is much more common to have them delivered to a central location for pick-up. You can also purchase bees from locals, but this option can put you at risk of transferring disease. Be very careful in choosing your bee source.

The best time to install your package is late afternoon or evening. Fifteen minutes before you are ready to move your bees to the hive, it is a good idea to mist a little bee feed syrup on the screen of the cage; this quiets them down inside the hive.

Before you even think about installing your bees, be sure to suit up properly. Installing the package into your hive is quite intimidating as you're staring at a box full of 10,000 active bees for the first time!

You should start the process by removing three to four frames from your hive and place them to the side. (You will put these frames back into your hive after your bees are safely installed.) Because you will need to feed your bees on a regular basis when first establishing your hive, you also should put an in-hive feeder into the hive now. The package will contain a can of syrup that you can use for the initial feeding, but you'll need a supply of high-quality beefeed on hand for subsequent feeding.

To prepare the package for installation, you should first pry up the package's feeder can. Gently tap the box on the ground a few times to shift the bees toward the bottom of the cage. Be careful to prevent the queen cage from slipping down into the cage.

With a firm grip, remove the feeder can and pull the queen cage out of the package. Immediately replace the feeder can over the hole on the top of the cage to prevent the bees from getting out before you are ready for them.

With the queen cage removed, quickly inspect your queen to make sure she is alive and in good condition. Carefully remove the cork and insert the candy at one end of the queen cage. The bees will release the queen from the cage by eating the candy. (Because she is a stranger to the colony, this gives the worker bees time to become accustomed to her, increasing her chance of survival.)

Place the queen cage securely between two frames in the center of the hive body. You can staple the strap into place to keep the cage from falling.

Again, tap the package lightly a few times to move the bees to the bottom of the cage. Remove the can lid and gently invert the cage over the hive to shake the bees from the package. They will pour over the tops of the frames and into the void space.

Once you have shaken the bees from the package, place the package on the ground in front of the hive so that the few remaining bees can fly out and into the hive through the entrance.

Next, you should gently replace the frames you removed earlier. You may need to slide the bees very gently down between the frames and into the brood box. You can then close up your hive with the inner and top cover. Be patient and give them a little time and guidance. They want to be near their queen.

Congratulations! You have just established your first honeybee colony!

In about a week, you will want to inspect the hive to make sure the queen has been released and is still alive. Remove the queen cage during this first inspection. Then, you'll want to check the hive every other week to make sure the feeder is full, the queen is laying in a healthy brood pattern, and the workers are building up a good honey store.

But other than that, your job is now very simple: Leave them alone. They will begin making themselves at home inside their hive by building comb and storing honey almost immediately. Resist the temptation to inspect them too frequently at this stage. They can be made to feel threatened by frequent disturbance.

CARING FOR YOUR *Colony*

O nce you have installed your bees in their brood box, they will naturally start building a community. Your job in these first weeks is simple: Give them a comfortable home and space. While they're getting settled, you don't want to check or feed them more than every other week. When you do inspect the hive and refill their feeder, be certain to use smoke to keep yourself safe and the bees calm and docile.

Within two months, your colony will have filled most of the frames in the first brood box and will be looking to expand. At this point, you'll want to add an additional brood box on top of the first. This box will contain additional brood and the honey stores the colony will need to make it through the winter. To draw the bees up into their new space, remember to move the inside hive feeder to the second box.

When you notice white wax capped-off cells on nine of the ten honeycomb frames in the second brood box, you should insert a queen excluder to keep the queen in place.

Next, add your first honey super. These medium-sized boxes will hold your honey harvest. A healthy hive will start to fill the honey super in as little as two weeks. If you prefer to harvest more honey, simply add additional supers as they fill. A healthy hive can easily accommodate as many as six honey supers.

INSPECTIONS

Inspections, which should take place no more than every other week, are primarily intended to determine the health and productivity of the colony. Always begin your inspection of the hive by smoking the bees and removing the first frame or the frame closest to the outer wall. Insert the end of your hive tool between the first and second frames, near one end of the frame's top bar. Twist the tool to separate the frames from each other. Repeat this motion at the opposite end of the top bar. Using both hands, pick up the first frame by the end bars and gently rest it on the ground, leaning it vertically up against the hive. If you have a frame rest, use it to temporarily store the frame. Now you have room to manipulate the other frames and remove them without the risk of injuring any bees.

Holding and inspecting an individual frame with the sun behind you, shining over your shoulder and onto the frame illuminates details deep in the cells and helps you to better see eggs and small larvae. Turn the frame slowly to inspect both sides. Work your way through all ten frames in this manner. When you're done looking at each frame, always return it snugly against the frame previously inspected.

Pay attention to the bees. A few minutes into your inspection, you may notice that the bees have lined up between the top bars with their heads in a row between the frames. Kind of cute, aren't they? They're actually watching you and are getting ready to defend their hive. That's your signal to soothe them with a few puffs of smoke so you can safely continue with your inspection.

INSPECTION CHECKLIST

Each time you visit your hive, you should keep in mind the following checklist of the things you should inspect.

CHECK UP ON YOUR QUEEN:

You want to be looking for indications that the queen is alive and well and laying eggs. Rather than trying to inspect the queen herself, however, you should be looking for her eggs. Although they're tiny, finding the eggs is much easier than locating a single queen in a hive of 60,000 bees.

You also should inspect the hive's brood pattern. A tight, compact brood pattern is indicative of a good, healthy queen. Conversely, a spotty brood pattern (many empty cells with only occasional cells of eggs, larvae, or capped brood) is an indication that you have an old or sick queen and may need to replace her. Look for a u-shaped ring of honey cells above brood cells.

CONFIRM SUFFICIENT FOODSTUFFS:
You should learn to identify the different food materials collected by your bees and stored in their cells. For example, they pack pollen in some of the cells that can come in many different colors: orange, yellow, brown, gray, blue, and so on. Bees also fill cells with liquid that can be either nectar or water.

LOOK FOR SIGNS OF DISEASE:

There are many diseases that can strike your colony. It is important to understand the signs of parasites, fungus, mold, infection and other health risks. We explore these concerns in more depth in the next chapter.

ASSESS CAPPED CELLS/FULL COMB:
When you see half an inch of consistent wax built out across your combs and signs of brood growth on a majority of frames, it's time to give your bees more space. You should be able to see the signs of a built-out honeycomb even before you lift out the frame. The solution is simple: You need to add an additional honey super.

ZERO IN ON EXCESS WAX AND PROPOLIS: You should use your hive tool to scrape away excess wax and propolis that can build up on the edges of frames and inside cover. The wax can be left near the entrance of the hive for the bees to recycle.

Once you're done with your inspection, you should gently replace the inner and outer cover, and leave your bees to work their magic undisturbed for another two weeks.

BEEKEEPER'S CALENDAR

The way that you care for your hive during the year will vary slightly depending on the season. This beekeeper's calendar offers some general guidelines for what you should be doing when. Directions are based on an elevation of 4,000 feet. Climates in southern and northern states might be a month earlier or later.

JANUARY

During winter, your bees will tend to collect into a tight ball to maintain the warmth they need to survive. You should monitor their stores and feed them as necessary.

FEBRUARY

Toward the end of the month, the queen will be stimulated to start egg laying. The colony will need to raise the temperature for the new pupae, and they do this by consuming more food. With food stores probably already low and still no nectar coming in, depending on your area and climate, you should be especially vigilant about feeding. That said, the queen won't start laying eggs if it's still too cold outside; check the hive only if outside temperatures are above 55°F.

MARCH

With weather warming and early pollen being brought in, bees will begin feeding protein to older larvae and young bees. You should be watching the main food stores during this time. Toward the end of the month, you'll want to light your smoker and delve into the brood chamber for your first inspection of the year. Check for damaged frames and signs of water in the hive, and check the brood pattern and larva for diseases. Again, you should only begin your inspections when outside temperatures reach 55°F.

APRIL

Spring is a busy time for bees as temperatures rise. Drones leave the hive with their drone congregation, the queen increases her egg-laying to around 1,000 a day, and the adult population rises to approximately 30,000 to 40,000. Depending on region, you may need to give your growing colony more room by adding an extra box. You should inspect your colony about every 10 days.

MAY

With nectar starting to flow from fruit trees, the queen gets an extra boost in her egg-laying prowess. Consequently, by about the middle of the month, you should expect to see the brood chamber full of brood and honey stores. This also may be the time you see a supersedure within your hive, as a new queen is grown and a portion of the colony splits off with the original queen to create a new hive.

JUNE

If your colony has experienced a supersedure, your bees have a lot of additional work to do with a virgin queen in a caste to ensure the colony grows to a sufficient size to gather stores to survive the winter. As the main colony settles down, the remaining bees give their attention to the main honey flow. Summertime is "all systems go" for the collection of pollen and nectar, as well as for the production of honey. You might want to consider splitting your hive at this time.

JULY

As summer peaks at the end of the month, your colony should have settled with a mated laying queen, and most of the worker bees will be out gathering nectar. Your job is to give the colony plenty of room by adding empty supers to hold all of their honey.

AUGUST

Toward the end of the month, your colony will begin to shut down and go into protective mode as your bees guard their precious stores of honey. Look for signs of varroa mite and treat it right away, so that the colony can go into winter mode in the best possible health. This also is the time to harvest your share of the colony's honey. Wait for a warm evening when your bees are calm to extract honey, and focus your efforts on a majority of combs with wax capping.

SEPTEMBER

Your colony's population will start to drop as drones are evicted from the hive. The queen will slow her rate of egg production, further dropping the colony's numbers. You should be vigilant about inspecting the health of your bees during this time, and treating them as necessary for disease and parasites. You also should take an inventory of your bees' food store to ensure they have two boxes of capped honey and brood, which they will need to survive the winter. You can make up for any shortfalls with beefeed, which you should administer by way of in-hive feeders.

OCTOBER

On warm fall days, the workers may still bring some pollen and nectar to the hive. However, you should install entrance reducers to keep out other bees and invading insects, as well as to keep out frigid autumn winds. Entrance reducers also will help your bees in the spring, as stronger colonies may try to steal hives and honey from your relatively weaker colony.

NOVEMBER

Your colony will get through the winter with the queen and about 10,000 workers. Some beekeepers wrap up their hives for the winter, but we recommend caution, as wrapping could trap your bees inside the hive. You should conduct external inspections for damage from wind and woodpeckers; this also is a good month for cleaning and repairing equipment that will be used next year.

DECEMBER

The bees in your colony will start to gather into a cluster around their queen. They consume honey to keep their wings constantly vibrating, which keeps the temperature just warm enough for survival. You should watch your in-hive feeder to ensure they have a sufficient food supply to get them through the winter. However, to ensure you don't expose your bees to excessive cold, it's important that you open up your hive only if outside temperatures are above 55°F.

Protecting Your Bees FROM HARM

Once a hive has been placed and populated, it can hum along with minimal day-to-day interference. However, the one place where you make a life-or-death difference is in protecting your colony from infection and parasites. It is essential to become familiar with the signs of common bee diseases and how to prevent them in your apiary.

VARROA MITES: Varroa mites are parasites that suck the blood of adult and larval honeybees, weakening them and shortening their lifespan. Varroa mites also are carriers for a virus that is particularly damaging to the bees if they are exposed during their development. If your bees display signs such as missing or deformed legs or wings, it is possible your colony has become infected by varroa mites. Frequent inspections are the best preventative tool. Varroa mites can be seen with the naked eye, as small red or brown spots on the bee's thorax. Varroa mites are especially attracted to the drone brood, so pay particular attention to looking for them there. Mite Away seems to work best for treating for varroa mites, but no medicine should be used during honey flow or if honey will be shortly harvested for human consumption. Follow the manufacturer's instructions.

ACARINE (TRACHEAL) MITES: Acarine mites are small parasitic mites that infect the airway of the honeybee. Diagnosis for tracheal mites generally involves dissecting a bee from the hive and examining it under a microscope. Mature female acarine mites will leave the bee's airway and climb out on a hair of the bee, where they wait until they can transfer to a young bee. Once on the next bee, they will move into the bee's airways and begin laying eggs. Acarine mites are commonly controlled with grease patties (typically made from 1 part vegetable shortening mixed with 3–4 parts powdered sugar), which are placed on the top bars of the hive. As the bees eat the sugar, they pick up traces of shortening that disrupts the mite's ability to identify a young bee. Menthol that is either allowed to vaporize from crystal form or is mixed into the grease patties also is often used to treat for acarine mites. If you have frequent infestations of acarine mites, you should consider investing in a resistant hybrid bee known as the Buckfast bee to combat this disease.

NOSEMA: Nosema is an infection associated with black queen cell virus. Nosema normally is seen only when the bees can't leave the hive to eliminate waste. When the bees are unable to take cleansing flights, they can develop dysentery. If your hive develops nosema, you should increase the ventilation of the hive. Some beekeepers also treat hives with antibiotics such as fumagillin. Nosema also can be prevented or minimized by removing much of the honey from the beehive, then feeding the bees on bee feed in the late fall. The refined sugars in bee feed have lower ash content than flower nectar, which reduces the risk of dysentery.

SMALL HIVE BEETLE: When colonies become infested with a small, dark beetle known as the small hive beetle, they will eventually leave their hive. Generally, the same controls that beekeepers put in place to prevent ants from climbing into the hive are believed to be effective against the hive beetle as well. Some beekeepers also use diatomaceous earth around the hive as a way to disrupt the beetle's lifecycle, or place cardboard filled with the chemical Fipronil around the hive. The standard corrugations of cardboard are large enough for small hive beetles to enter, but small enough that honeybees can't enter. Alternative controls such as oil-based traps also are available.

WAX MOTH: Wax moths do not attack bees directly, but feed on the wax used by bees to build their honeycomb. The destruction of the comb will spill or contaminate stored honey, and may kill bee larvae. When you find moth-damaged combs, you should scrape them out; they will be replaced by the bees. Because freezing temperatures kill wax moth larvae and eggs, they are usually not a problem for beekeepers in the northern United States and Canada, and storing hives in unheated sheds or barns in higher latitudes is the only control needed. Also, a strong hive generally needs no treatment to control wax moths; the bees themselves will kill and clean out the moth larvae and webs. If wax moths become a problem that a hive cannot deal with, they can be controlled chemically with moth crystals or urinal disks. The use of naphthalene mothballs is discouraged, however, because it accumulates in the wax and can kill bees or contaminate honey stores. If chemical methods are used, the combs should be aired out for several days before being returned to the bees.

AMERICAN AND EUROPEAN FOULBROOD (AFB, EFB):

Caused by spore-forming larvae, foulbrood is the most widespread and destructive bee brood disease. Larvae up to three days old become infected by ingesting spores present in their food. Infected larvae darken and normally die after their cell is sealed. Although this disease only affects bee larvae, not adult bees, it is highly infectious and deadly to the brood. Test kits are available, meaning that you should confirm foulbrood before treating for it chemically using oxytetracycline hydrochloride. That said, you should not use any disinfecting medicine during honey flow or if you plan on harvest honey in the hive at a future date. Foulbrood spores are especially dangerous because they can remain viable for more than 40 years in honey and beekeeping equipment. Each dead larva may contain as many as 100 million spores. If you are unable to contain a foulbrood outbreak, you may need to destroy your colony and possibly your hive.

CHALKBROOD:

Chalkbrood is caused by a fungal disease that infests the gut of the larva. The fungus, known as Ascosphaera apis, will compete with the larva for food, ultimately causing it to starve. The fungus will then go on to consume the rest of the larva's body, causing it to appear white and chalky. Chalkbrood is most commonly visible during the rainy spring season. Infected hives can generally be treated by simply increasing ventilation.

CHILLED BROOD:

This is not actually a disease, but rather the result of a beekeeper opening the hive when temperatures are too low. The brood must be kept warm at all times; nurse bees will cluster over the brood to keep larvae at the right temperature. When a beekeeper who opens the hive for too long prevents nurse bees from clustering on the frame, the brood can become chilled, deforming or even killing some of the bees. When inspecting your hives, you need to always be conscientious, quick, and aware of the ambient temperature. Never open your hive when the outside temperature is below 55°F. Chilled brood also can be caused by a sudden drop in the outside temperature in the spring as the colony is rapidly building up its numbers, catching the nurse bees off guard.

PESTICIDE POISONING: Honeybees are vulnerable to many of the chemicals used to kill other common pests. Because bees forage up to several miles from the hive, they may fly into areas actively being sprayed or collect pollen from contaminated flowers. Pesticide toxicity can take as long as two days to become evident. Large and sudden numbers of dead bees in front of the hive can be the result of pesticide. Because pesticide label instructions must be followed precisely to protect pollinators, your single best defense is simply to increase awareness of responsible pesticide use in your neighborhood.

COLONY COLLAPSE DISORDER (CCD):

Colony collapse disorder is a new and poorly understood phenomenon in which worker bees abruptly disappear from the hive and never return. Initial hypotheses for the causes of CCD vary widely and include environmental stresses, malnutrition, pathogens, mites, or a class of pesticides known as neonicotinoids. In 2010, U.S. researchers identified a co-infection of invertebrate iridescent virus type 6 and Nosema ceranae in all CCD colonies sampled. That said, there is still no consensus regarding what causes colony collapse disorder, and no solution to this problem except burning the hive once the colony has been infected.

The diseases described above are just the most common ones to strike bee colonies. There are other diseases that can threaten bee colonies and alternate methods to prevent them. The best way to be aware of local outbreaks and infestations is to be connected to your local beekeeping community. Your local beekeeping community also can share with you successful methods for dealing with these challenges.

HARVESTING YOUR
Sweet Reward

The wait for your first honey harvest can seem like a long one, as a new colony needs at least a few months to build up sufficient honey stores. With some new colonies, you may be forced to wait a full season before your bees have surplus honey to share.

Once your bees start building up honey in their honey supers, you can start monitoring the hive for the telltale signs that it's safe to begin harvesting your sweet reward.

You should look for fully built-out frames in which about 80 percent of the cells are enclosed beneath white cappings, as this is a good indicator that it is time to extract the honey or add a new supers to promote additional production. You can add additional supers throughout the season, but in colder climates, you should extract your honey before daytime temperatures drop below 80 degrees. As soon as the weather turns cooler, bees will start to consume their honey to prepare for the winter.

When you harvest honey, it's important to remember that you must leave enough honey supply for the colony to use throughout the winter. At minimum, you must not touch the honey in the large super beneath the queen excluder; your bees will die without this essential food source. By respecting their harvest, you are ensuring year after year of your own sweet reward.

When you're ready to harvest your honey, you should select a warm day while your bees are out foraging. Before you begin the honey extraction process, you should set aside an enclosed space to work in and gather all of the necessary tools for the job. Keep in mind that the sweet scent of honey will naturally attract every bee in the area, so you want to choose an enclosed space like a shed, barn, or closed garage that will be safer for you and minimize interruptions.

When you suit up, you should wear complete protective clothing and have your smoker handy at all times. Even the most docile bees can become agitated when their honey is being stolen. You also must be prepared to lift heavy honey frames that can weigh up to 50 pounds when full. You may want to use a wagon or wheelbarrow to transport the frames away from the hive.

You should assemble the following tools:

- Hive tool
- Protective clothing and gloves
- Smoker
- A wagon or other way to transport supers
- Bee brush
- Uncapping knife and scratcher
- Food grade buckets with lids
- Cheesecloth or metal sieve
- Extractor
- Jars and labels

You should start by smoking the hive, then wait for two or three minutes. When the bees seem sufficiently calm, open the hive and use your hive tool to loosen the frames. Using frame grips, carefully remove the honeycombs and gently return the bees to the hive with a bee brush.

During the harvesting process, an extra honey super box will come in handy to hold the heavy honeycomb. You can put the box into a wagon or well-balanced wheelbarrow for easy transporting to your extraction area.

When you move your combs to your enclosed extraction area, you will want to work quickly. Extracting honey is easier when the honey is still warm from the beehive, as it flows much more freely.

You should hold the frame over a large, clean container to catch honey drips (like a cookie sheet). With an angle knife or scratcher, you should work to expose the honey within the cells of each frame. A gentle side-to-side slicing motion works best, like slicing bread. Start a quarter of the way from the bottom of the comb, slicing upward. Remember to keep your other hand away from the knife in case your knife slips. When you reach the lower area

of the frame, thrust your knife downward thrust to uncap those cells. Use the scratcher to puncture any cells that remain capped. Then, turn over the frame and repeat this process on the opposite side.

Next, you should place the uncapped frames into an extractor to spin the honey out of the comb and into a holding tank. Remember that extractors aren't the lazy way out; they are the fastest, easiest, and most efficient way to harvest the maximum amount of honey from each comb. Plus, the extractor keeps the comb intact, allowing you to return it to the hive for the bees to refill.

Once you've uncapped enough frames to fill your extractor, affix the lid to the extractor and start cranking. Spin slowly at first, building some speed as you progress, but never apply maximum speeds, as the extreme centrifugal force may damage the delicate wax comb. After spinning for ten to fifteen minutes, turn over the frames to extract the opposite sides. After another ten to fifteen minutes of spinning, the comb should be empty and you can return the harvested frames to the honey super.

As the honey tank of your extractor fills, it will become more difficult to spin, which is your signal to open the valve (honey gate) at the bottom of the extractor. The honey will flow out and place a honey sieve or filter on a honey bucket with a gate (or spigot) that makes it simple to fill and seal bottles with minimal mess. You can expect to harvest approximately 30 pounds of honey from each complete honey super. Once your bottles are filled, wipe them down with a warm, wet cloth and label them with the extraction date.

If you are harvesting your honey in an early summer extraction in July or August, you should return the emptied frames to the top of your hive, as the bees will have plenty of time to refill them before fall and you can squeeze in a second honey harvest. If you are harvesting in September, you should instead put the empty box near the hive for two days to allow the bees to clean away the remaining honey. Then, store the empty super until next season (remove honey super from yard at night time).

You should wait until dusk, when your bees are back in their hive, to begin cleaning your tools and extractor. Plain hot water is the best cleaning solution, as it effectively removes both honey and wax from all surfaces and tools. After you've washed all of your tools down thoroughly, store them in a sealed container until next year. Before storing your extractor, treat the bearing at the base of the extraction chamber with food-safe oil; then, cover the extractor with a tarp.

AFTERWORD

The benefits of beekeeping are many—relaxation and self-reliance, lending support to your flowers and fruit trees, and the personal satisfaction of knowing you are doing your part to conserve endangered pollinators. And let's not forget the sweetest reward of all—fresh, organic, home-crafted honey.

One of the best features about honey is that it never goes bad. Stored honey will naturally crystallize over time, but that does not mean you should throw it out. To restore your honey to its liquid form, simply warm your honey jars in a large bowl of hot water. Just be cautious about transferring cold jars into hot water, as they can crack. (You should always bring any jar to room temperature before exposing it to heat.)

If you're still feeling conscientious or hesitant about raising a bee colony, remember that bees are comfortable almost anywhere, even on balconies in the middle of big cities. And despite what popular culture might suggest, beekeeping is safe and simple in any setting, as long as you have the proper tools and maintenance. Even your children and pets can co-exist peacefully with your bees.

That said, when it comes to raising honeybees, there's a lot to learn about keeping your colony safe and secure. This book is designed to answer your initial questions and help you with basic considerations, but it really should be the beginning of your learning and discovery process. Volumes of information exist online, at your local library, and in many informative books (HLH will have additional beekeeping books to help as you grow into this wonderful hobby). As you become more involved in beekeeping, you'll find you want to just keep learning more and more.

ONE FINAL NOTE:

We have written this book for backyard beekeepers—folks who use the honey they harvest for personal use. There are strict regulations in many areas of the country regarding selling honey, so if you're thinking about selling honey products, it is essential that you research your local laws.

ABOUT HARVEST LANE HONEY

The legacy of beekeeping runs deep at Harvest Lane Honey. In the early twentieth century, Ollie Justice supported an apiary in the rugged climate of western Utah. His hives supplemented his family's sugar needs during wartime rationing.

Three generations later, Jason and Mindy Waite noticed their fruit and vegetable garden yields were diminishing. Together with their parents, Mike and Rhonda Wells, they reawakened their great-grandfather's passion for bees as a way of supporting their backyard harvest. In just one season, they were all hooked.

With a background in drafting and woodworking, Jason soon became frustrated with flimsy hives that just met industry standards. He began engineering a better hive based on the classic Langstroth design and high-grade construction standards.

HARVEST LANE HONEY
H.L.H.™

www.harvestlanehoney.com

He needed a sturdy hive able to withstand the hot summers, high winds, and harsh winters that are all too common in the Wasatch Mountains. He wanted hives that were sturdy enough to handle with ease. He also wanted safe, healthy hive that supported healthier bee colonies.

The resulting hives were a marked improvement on the original. Built from sturdy, ¾" pine and featuring fully encased tops built to withstand 3000 pounds and drain away rain and snow with a custom drip-edge, the Harvest Lane Honey hive is built to endure whatever nature throws their way. With two coats of high quality exterior paint and high quality vented bottom boards, they are moisture resistant to protect the bees.

Soon, the word got out, and local beekeepers were interested in hives that could stand up to their unique climate. What was initially a hobby soon became a family business. Since 2000, Harvest Lane Honey has grown from a small storefront to a significant retail brand available nationwide.

GLOSSARY OF BEEKEEPING TERMS

A

ABDOMEN
Segmented posterior part of bee containing heart, honey, stomach, intestines, reproductive organs, and sting.

ACARAPIS WOODI
Scientific name of acarine mite, which infests tracheae of bees.

ACARINE DISEASE
Condition caused by Acarapis woodi.

ACID BOARD (ALSO FUME BOARD)
A rimmed hive cover containing a pad of absorbent material into which benzadehyde or butyric anhydride (bee repellents) is poured. Used to remove bees from honey supers.

AHB
Africanized honeybee.

ALIGHTING BOARD
Extended entrance of beehive on which incoming bees land.

ALLELE
One of a pair or series of alternative genes that can occur at a given point on a chromosome.

AMERICAN FOUL BROOD (AFB)
Contagious disease of bee larvae caused by Bacillus larvae.

AMM (APIS MILLIFERA)
The European dark bee (Apis melliferaj) was domesticated in modern times, and taken to North America in colonial times. These small, dark-colored honey bees are sometimes called the German black bee, although they occurred originally from Britain to eastern Central Europe.

ANTENNAE
Slender jointed feelers, which bear certain sense organs, on head of insects.

ANTHER
Part of plant that develops and contains pollen.

APIARIST
Beekeeper.

APIARY
Group of bee colonies kept in one location (bee yard).

APICULTURE

The science and art of studying and using honey bees for man's benefit.

APIS CERANA

Scientific name of the Eastern honey bee, the honey producer of South Asia, also called Apis indica.

APIS DORSATA

Scientific name for the large honey bee of Asia which builds open air nests of single comb suspended from tree branches, rocky ledges, etc.

APIS FLOREA

Scientific name for the small honey bee of Asia.

APIS MELLIFERA

Scientific name of the Western honey bee.

APIS

The genus to which the honey bee belongs.

ARTIFICIAL INSEMINATION

(See Instrumental Insemination.)

AUTOPOLLINATION

The automatic transfer of pollen from anthers to stigma within a flower as it opens.

B

BACILLUS LARVAE

Bacterial organism causing American foulbrood.

BALLING A QUEEN

Clustering around unacceptable queen by worker bees to form a tight ball; usually queen dies or is killed in this way.

BEE BREAD

Pollen stored in cells of the comb and used by bees for food.

BEE DANCE

Anthropomorphic term for one of several physical maneuvers conducted within a bee colony; it has very inaccurate correlations relative to a forager's flight experience in the field (distance and direction of the site visited), but odor on the dancer's body appears to be the means of communication that recruits use to find the same nectar or pollen source.

BEE ESCAPE

Device to let bees pass in only one direction; usually inserted between honey supers and brood chambers, for removal of bees from honey supers.

BEE GUM
Usually hollow log hive.

BEE LOUSE
Relatively harmless insect that gets on honey bees, but larvae can damage honeycomb; scientific name is Braula coeca.

BEE LUST
An insatiable covetousness of more bees, hives, and/or beekeeping paraphernalia that leads one to amass more than they can possibly maintain or has the knowledge to manage.

BEE METAMORPHOSIS
The transformation of the bee from egg to larva to pupa and finally to the adult stage.

BEE MOTH
(See Wax Moth.)

BEE PARALYSIS
An adult bee disease of chronic and acute type caused by different viruses.

BEE SPACE
A space (1/4- to 3/8-inch) big enough to permit free passage for a bee but too small to encourage comb building. Leaving bee space between parallel beeswax combs and between the outer comb and the hive walls is the basic principle of hive construction.

BEE TREE
A hollow tree occupied by a colony of bees.

BEE VEIL
(See Veil.)

BEE VENOM
Poison injected by bee sting.

BEE YARD
(See Apiary.)

BEEHIVE
Domicile prepared for colony of honey bees.

BEESWAX
Wax secreted from glands on the underside of bee abdomen; molded by bees to form honeycomb.

BOARDMAN FEEDER
A small, plastic feeder placed at the hive entrance and holding an inverted pint or quart glass jar of sugar syrup.

BOTTOM BOARD
Floor of beehive.

BRACE COMB
Section of comb built between and attached to other combs.

BRAULA COECA
(See Bee Louse.)

BROOD CHAMBER
The area of the hive where the brood is reared; usually the lowermost hive bodies.

BROOD COMB

Wax comb from brood chamber of hive containing brood.

BROOD NEST

Area of hive where bees are densely clustered and brood is reared.

BROOD

Immature or developing stages of bees; includes eggs, larvae (unsealed brood), and pupae (sealed brood).

BT

Bacillus thuringiensis. Used for controlling wax moths. Bacillus thuringiensis aizawai strain NB200 is a part of a large group of bacteria, Bacillus thuringiensis (Bt), that occur naturally in soil. These bacteria are toxic to certain species of insects and can be used as an insecticides. Once ingested by larvae, Bt bacteria release a toxic protein into the insect digestive system. This protein causes death by attaching to the gut, eventually rupturing it. Different strains of Bt are toxic to specific groups of insects. Bacillus thuringiensis aizawai strain NB200 is known to be toxic to numerous species of moths, including many pests of agricultural crops.

BURR COMB

Comb built out of place, between movable frames or between the hive bodies.

C

CAPPED BROOD

Brood (either last larval stage or pupal stage) that has been capped over in its cell.

CAPPED HONEY

Cells full of honey, closed or capped with beeswax.

CAPPINGS SPINNER

A centrifuge with wire-screened baskets used to separate honey from wax.

CAPPINGS

Beeswax covering of cells of honey which are removed before extracting.

CARNIOLAN BEES

A race of honey bees which originated in the southern part of the Austrian Alps and northern Yugoslavia.

CASTES

The three types of individual bees (workers, drones, and queen) that comprise the adult population of a bee colony.

CAUCASIAN BEES

A race of honey bees native to the high valleys of the Central Caucasus.

CELL CUP

Initially constructed base of queen cell; also made artificially for queen rearing.

CELL

The six-sided compartment of a honeycomb, used to raise brood or to store honey and pollen. Worker cells approximate five to the linear inch, drone cells are larger averaging about four to the linear inch.

CHECKERBOARDING

A management technique to prevent swarming.

CHILLED BROOD

Brood that has died because of chilling. It can be a result of mistreatment of the bees by the beekeeper. It also can be caused by a pesticide hit that primarily kills off the adult population, or by a sudden drop in temperature during rapid spring buildup. The brood must be kept warm at all times; nurse bees will cluster over the brood to keep it at the right temperature. When a beekeeper opens the hive (to inspect, remove honey, check the queen, or just to look) and prevents the nurse bees from clustering on the frame for too long, the brood can become chilled, deforming or even killing some of the bees.

CHROMOSOMES

The structures in a cell that carry the genes.

CHUNK HONEY

A jar of honey containing both liquid (extracted) honey and a piece of comb with honey.

CLEANSING FLIGHT

Flight bees take after days of confinement, during which they void their feces.

CLIPPED QUEEN

Queen whose wing (or wings) has been clipped for identification purposes.

CLUSTER

Loosely, any group of bees that forms a relatively compact aggregation. A winter cluster is composed of all the bees in the colony huddled as closely together as necessary to maintain the required temperature. As the ambient temperature increases, the cluster expands until it loses its identity but it will reappear if the temperature drops.

COLONY COLLAPSE DISORDER (CCD)

A phenomenon in which worker bees from a beehive or European honey bee colony abruptly disappear. While such disappearances have occurred throughout the history of apiculture, the term colony collapse disorder was first

applied to a drastic rise in the number of disappearances of Western honey bee colonies in North America in late 2006. The cause or causes of the syndrome are not yet fully understood.

COLONY
Social community of several thousand worker bees, usually containing one queen, with or without drones. (See Social Insects.)

COMB FOUNDATION
Thin sheet of beeswax impressed by mill to form bases of cells; some foundation also is made of plastic and metal.

COMB HONEY
Honey marketed and eaten in the comb.

COMB
(See Honeycomb.)

CORBICULA
(See Pollen Basket.)

CREAMED HONEY
Honey made to crystallize smoothly by seeding with 10 percent crystallized honey and storing at about 57°F.

CROSS POLLINATION
Transfer of pollen between plants which are not of identical genetic material.

CRYSTALLIZED HONEY
Honey hardened by formation of dextrose-hydrate crystals. Can be reliquefied by gentle heat.

CUT COMB HONEY
Comb honey cut into appropriate sizes and packed in plastic.

D

DEARTH
Severe to total lack of availability, usually in reference to nectar and/or pollen.

DEMAREE
Method of swarm control, by which queen is separated from most of brood; devised by man of that name.

DEXTROSE
Also known as glucose; one of principal sugars of honey.

DIASTASE
Enzyme that aids in converting starch to sugar.

DIPLOID
An organism or cell with two sets of chromosomes, for example, worker and queen honey bees.

DISAPPEARING DISEASE
A condition in which colonies become weak from causes which are not readily identifiable.

DIVISION BOARD
Flat board used to separate two colonies or colony into two parts.

DRAWN COMB
Comb having the cells built out (drawn) by honey bees from a sheet of foundation. Cells are about 1/2-inch deep.

DRIFT
Movement of bees from their original hive into a neighboring hive frequent with drones and surprisingly common with workers.

DRONE COMB
Comb with about four cells to the inch and in which drones are reared.

DRONE CONGREGATION AREA (DCA)
An area where many drones from surrounding colonies gather to mate with queens during their nuptial flights.

DRONE LAYER
A queen which lays only unfertilized eggs which always develop into drones. Results from improperly or non-mated queen or an older queen who has run out of sperm.

DWINDLING
Rapid or unusual depletion of hive population, usually in the spring.

DYSENTERY
The discharge of fecal matter by adult bees within the hive. Commonly contributing conditions are nosema disease, excess moisture in the hive, starvation conditions, and low quality food. Tan, brown, or black fecal smears on combs or outside of hive indicate such a problem.

ESCAPE BOARD
Board with one or more bee escapes on it to permit bees to pass one way. Used to empty one or more supers of bees.

EUROPEAN FOULBROOD (EFB)
Brood disease of bees caused by Streptococcus pluton and possibly associated organisms.

EXTRACTED HONEY
Honey removed from the comb by centrifugal motion (in a special machine called an extractor) and marketed in the liquid form.

EXTRACTOR
Machine that rotates honeycombs at sufficient speed to remove honey from them.

FANNING

Worker bees fan the hive by directing airflow into the hive or out of the hive depending on need, sometimes cooling it with evaporated water brought by water carrier bees.

FESTOON

A unique cluster of bees that link themselves together by their tarsi (feet) in a loose network between combs in a hive. Normally, these are aggregates of wax-producing bees.

FGMO

Food Grade Mineral Oil. Has been used as an alternative treatment for honey bee mites.

FIELD BEES

Those bees in the hive who are mature enough to fly from the hive on foraging missions; also termed forager bees.

FOLLOWER BOARD

A board anywhere from 3/4" to 1/4" thick, plywood or other material, cut to the size of your frames (deep, med or shallow). A simple divider that acts like a movable hive side, allowing you to create any interior size needed.

FOOD CHAMBER

Hive body containing honey provided particularly for overwintering bees.

FOUNDATION

(See Comb Foundation.)

FRAME

Rectangular, wooden honeycomb supports, suspended by top bars within hive bodies.

FRUCTOSE

(See Levulose.)

FULL SISTERS

Queen or worker bees produced by a single queen and sired by different drones that are related to each other as brothers (used in bee breeding).

FUMAGILLIN

Antibiotic given bees to control nosema disease.

FUME BOARD

(See Acid Board.)

G

GALLERIA MELLONELLA
Scientific name of greater wax moth, whose larvae destroy honeycomb.

GAMETE
A male or a female reproductive cell (egg or sperm).

GENE POOL
The genetic base available to bee breeders for stock improvement.

GENE
A unit of inheritance located at a specific location in a chromosome.

GERMPLASM
All the hereditary material that can potentially contribute to the production of new individuals.

GIANT BEE
(See Apis Dorsata.)

GLUCOSE
(See Dextrose.)

GRAFTING
The transfer of young larvae from worker cells to queen cups.

GRANULATED HONEY
(See Crystallized Honey.)

H

HALF SISTERS
Queen or worker bees produced by a single queen and sired by drones that are not related to each other.

HAPLOID
An organism or cell with one set of chromosomes; for example, drone bee.

HARVEST LANE HONEY
Source of the highest quality products, sound education and friendliest service for the backyard beekeeper. A totally awesome beekeeping supplier!

HBH
Honey-Bee-Healthy, an essential oil additive to honey bee feed to control varroa mites, tracheal mites and to reverse the parasitic mite syndrome (PMS) seen in colonies infested with varroa mites.

HOMOZYGOUS
The condition in which only one allele of a pair is present. Drones are homozygous at all loci.

HETEROSIS
Hybrid vigor.

HETEROZYGOUS
An organism with unlike members of any given pair or series of alleles (bee genetics).

HFCS
High Fructose Corn Syrup

HIVE STAND
A device that elevates the bottom board up off the ground.

HIVE TOOL
Metal tool for prying supers or frames apart.

HIVE
Man-constructed home for bees.

HMF (HYDROXYMETHYLFURFURAL)
An organic compound derived from dehydration of sugars.

HOFFMAN FRAME
Self-spacing wooden frame of type customarily used in Langstroth hives.

HOMOZYGOUS
An organism with identical members of any given pair or series of alleles.

HONEY BEE
Genus Apis, family Apidae, order Hymenoptera.

HONEY BOUND
When the brood nest is bounded or restricted by cells/comb filled with honey.

HONEY EXTRACTOR
(See Extractor.)

HONEY FLOW
Period when bees are collecting nectar from plants in plentiful amounts.

HONEY HOUSE
Building in which honey is extracted and handled.

HONEY PUMP
Pump for transferring liquid honey, usually from the extractor to storage tanks.

HONEY STOMACH (HONEY SAC)
An enlargement of the posterior end of the oesophagus in the bee abdomen. It is the sac in which the bee carries nectar from flower to hive.

HONEY SUMP
Temporary honey-holding area with baffles usually placed between the extractor and the honey pump; tends to hold back sizable pieces of wax and comb.

HONEY
Sweet, viscous fluid elaborated by bees from nectar obtained from plant nectaries, chiefly floral.

HONEYCOMB

Comb built by honey bees with hexagonal back-to-back cells on median midrib.

HONEYDEW

Sweet secretion from aphids and scale insects.

HOT ROOM

An insulated portion of a warehouse with radiant or forced air heating that can produce temperatures up to 100°F.

HYBRID

Offspring from two unrelated (usually inbred) lines.

HYMENOPTERA

Order to which all bees belong, as well as ants, wasps, and certain parasitic insects.

I

INBRED

A homozygous organism usually produced by inbreeding.

INBREEDING

Matings among related individuals.

INNER COVER

A cover used under the standard telescoping cover on a bee hive.

INSTRUMENTAL INSEMINATION

The act of depositing semen into the oviducts of a queen by the use of a man-made instrument.

INTEGRATED PEST MANAGEMENT (IPM)

A pest control strategy that uses a variety of complementary strategies including mechanical devices, physical devices, genetic, biological, cultural management, and chemical management. These methods are done in three stages prevention, observation, and intervention. It is an ecological approach with a main goal of significantly reducing or eliminating the use of pesticides while at the same time managing pest populations at an acceptable level.

INTRODUCING CAGE

Small wood and wire cage used to ship queens and also sometimes to release them into the colony.

INVERTASE

Enzyme produced by bees that speeds inversion of sucrose to glucose and fructose.

INVERTED OR INVERT SUGAR SYRUP

A mixture of glucose and fructose. It is obtained by splitting sucrose into its two components. Compared with its precursor sucrose, inverted sugar is sweeter and its products tend to stay moist and are less prone to crystallization. Inverted sugar is therefore valued by bakers, who refer to the syrup as 'trimoline' or 'invert syrup'.

IPM

(See Integrated Pest Management.)

ITALIAN BEES

A race or variety of honey bee which originated in Italy and has become widely dispersed and cross-bred with other races.

J

JUMBO HIVE

Hive 2-1/2 inches deeper than standard Langstroth hive.

L

LANGSTROTH FRAME

9-1/8- by 17-5/8-inch standard U.S. frame.

LANGSTROTH

A minister from Pennsylvania who patented the first hive incorporating bee space thus providing for removable frames. The modern hive frequently is termed the Langstroth hive and is a simplified version of similar dimensions as patented by Langstroth.

LARVA

Stage in life of bee between egg and pupa; "grub" stage.

LAYING WORKER

Worker bees which lay non-fertilized eggs producing only drones. They occur in hopelessly queenless colonies. Laying workers will lay multiple eggs per cell, have a spotty brood pattern, eggs laid on the sides of the cell or off center, and drone brood in worker sized cells.

LEVULOSE

Noncrystallizing sugar of honey which darkens readily if honey is overheated.

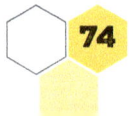

LINE BREEDING
Mating of selected members of successive generations among themselves in an effort to maintain or fix desirable characteristics.

LOCUS
A fixed position on a chromosome occupied by a given gene or one of its alleles.

M

MANDIBLES
Jaws of insects.

MATING FLIGHT
The flight of a virgin queen during which time she mates with 6–10 drones high in the air away from the apiary.

MEAD
A wine made with honey. If spices or herbs are added, the wine usually is termed metheglin.

METAMORPHOSIS
Changes of insect from egg to adult.

MIGRATORY BEEKEEPING
Movement of apiaries from one area to another to take advantage of honey flows from different crops.

MITE
(See Acarapis Woodi and Varroa Jacobsoni.)

MUTATION
A term used to describe both a sudden change in the alleles or chromosomes of an organism and the changed form itself as it persists.

N

NECTAR
A sweet secretion of flowers of various plants, some of which secrete enough to provide excess for the bees to store as honey.

NECTARIES
Special cells on plants from which nectar exudes.

NOSEMA DISEASE
Disease of bees caused by protozoan spore-forming parasite, Nosema apis.

NUCLEUS (NUKE, NUC)
A small colony of bees resulting from a colony division. Also, a queen-mating hive used by queen breeders.

NURSE BEES
Three-to 10-day-old adult bees that feed the larvae and perform other tasks in the hive.

O

OBSERVATION HIVE
Hive with glass sides so bees can be observed.

OCELLUS (OCELLI)
Simple eye(s) of bees.

ORIENTATION FLIGHTS
Short orienting flights taken by young bees, usually by large numbers at one time and during warm part of day.

P

PACKAGE BEES
A quantity of bees (2 to 5 lb) with or without a queen shipped in a wire and wood cage to start or boost colonies.

PARALYSIS
(See Bee Paralysis.)

PARTHENOGENESIS
Production of offspring from a virgin female.

PHEROMONES
Chemicals secreted by animals to convey information or to affect behavior of other animals of the same species.
(See Queen Substance.)

PISTIL
The combined stigma, style, and ovary of a flower.

PMS (PARASITIC MITE SYNDROME)
For years we have been seeing diseased bee larvae with symptoms resembling a cross between foulbrood and sacbrood. The USDA Beltsville Bee Lab has found these diseased larvae to be infected with one, or commonly several, viruses. This new disease seems to be limited to colonies infested with Varroa mites. Additionally, beekeepers have experienced bees disappearing completely from previously healthy colonies in the early fall. This situation is most likely associated with Varroa mites, viruses or a combination of both.

POLLEN BASKET
Area on hind leg of bee adapted for carrying pellets of pollen.

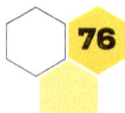

POLLEN CAKE
Cake of sugar, water, and pollen or pollen substitute, for bee feed.

POLLEN SUBSTITUTE
Mixture of water, sugar, and other material, such as soy flour, brewer's yeast, etc., used for bee feed.

POLLEN SUPPLEMENT
Pollen substitute added to natural pollen in a pollen cake.

POLLEN TRAP
Device which forces bees entering hive to walk through a 5-mesh screen, removing pollen pellets from their legs into a collecting tray.

POLLEN
Male reproductive cells of flowers collected and used by bees as food for rearing their young. It is the protein part of the diet. Frequently called bee bread when stored in cells in the colony.

POLLINATION
The transfer of pollen from the anthers of a flower to the stigma of that or another flower.

POLLINATOR
The agent which transfers pollen; e.g., a bee.

POLLINIZER
The plant source of pollen used for pollination; e.g., pollinizer varieties of apples and pears must be planted in order to produce a crop. Bees must carry the pollen from one variety to another.

PROBOSCIS
Mouth parts of bee for sucking up nectar, honey, or water.

PROPOLIS
A glue or resin collected from trees or other plants by bees; used to close holes and cover surfaces in the hive. Also called bee glue.

PUPA
Stage in life of developing bee after larva and before maturity.

Q

QUEEN CAGE CANDY
A special fondant made from Nulomoline, drivert, and glycerine; used to feed queen and attendant bees in queen cages.

QUEEN CELL
Cell in which queen develops.

QUEEN CUP
The beginnings of a queen cell in which the queen may lay a fertile egg to start the rearing of another queen.

QUEEN EXCLUDER

Device usually made of wood and wire, with opening 0.163 inch, to permit worker bees to pass through but excludes queens and drones. Used to restrict the queen to certain parts of the hive.

QUEEN SUBSTANCE

Pheromone material secreted from glands in the queen bee and transmitted throughout the colony by workers. It makes the workers aware of the presence of a queen.

QUEEN

Sexually developed female bee. The mother of all bees in the colony.

QUEENRIGHT

A colony of bees with a properly functioning queen.

R

RACE

Populations of bees, originally geographically isolated and somewhat adapted to specific regional conditions.

RENDERING WAX

Melting old combs and wax cappings and removing refuse to partially refine the beeswax. May be put through a wax press as part of the process.

REQUEEN

To replace a queen in a hive. Usually to replace an old queen with a young one.

RIPENING

Process whereby bees evaporate moisture from nectar and convert its sucrose to dextrose (glucose) and levulose (fructose), thus changing nectar into honey.

ROBBING

Bees steal honey from other hives. A common problem when nectar is not available in the field.

ROPINESS

Having the characteristic of sticky elasticity and stringing out when stirred and stretched.

ROYAL JELL

Glandular secretion of young worker bees used to feed the queen and young brood.

S

SAC BROOD
A fairly common virus disease of larvae, usually nonfatal to the colony.

SCALE
A dehydrated, dead larva shrunken to an elongated thin, flat chip at the bottom of a cell.

SCOUT BEES
Worker bees searching for nectar or other needs including suitable location for a swarm to nest.

SCREENED BOTTOM BOARD (SBB)
Acts as the floor of the hive, allowing mites to fall through to the ground instead of returning to the nest.

SEALED BROOD
Brood in pupal stage with cells sealed.

SELF-POLLINATION
The transfer of pollen from the anther to the stigma of the same flower or to flowers of the same plant or other plants of identical genetic material such as apple varieties, clones of wild blueberries, etc. (See Autopollination.)

SEPTICEMIA
Usually minor disease of adult bees caused by Pseudomonas apiseptica.

SMALL HIVE BEETLE (SHB)
The small hive beetle (Aethina tumida) can be a destructive pest of honey bee colonies, causing damage to comb, stored honey and pollen. If a beetle infestation is sufficiently heavy, they may cause bees to abandon their hive. The beetles can also be a pest of stored combs, and honey (in the comb) awaiting extraction. Beetle larvae may tunnel through combs of honey, feeding and defecating, causing discoloration and fermentation of the honey.

SKEP
A beehive, usually of straw and dome-shaped, that lacks movable frames.

SLATTED BOTTOM RACK
A ventilation board that fits between the bottom hive body and the bottom board (Langstroth Hive). It provides cluster space for bees, allows air circulation without allowing a direct draft on the brood, and helps prevent swarming.

SLUMGUM
A dark residue, consisting of brood cocoons and pollen, which is left after wax is rendered by the beekeeper.

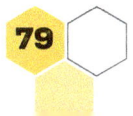

SMOKER

Device used to blow smoke on bees to reduce stinging.

SMR (SUPPRESS MITE REPRODUCTION)

Scientists at the Honey Bee Breeding Genetic & Physiology Laboratory (USDA, Agricultural Research Service) in Baton Rouge, Louisiana, have selected bees that are resistant to this [varroa] mite. The mechanism of resistance is a trait of the honey bee that suppresses mite reproduction (SMR). This trait prevents female mites from producing progeny. Because SMR is a trait rather than a stock, SMR genes can be added to any population of honey bees by using traditional breeding methods.

SOCIAL INSECTS

Insects which live in a family society, with parents and offspring sharing a common dwelling place and exhibiting some degree of mutual cooperation; e.g., honey bees, ants, termites.

SOLAR WAX MELTER

Glass-covered box in which wax combs are melted by sun's rays and wax is recovered in cake form.

SPERMATHECA SMALL SACLIKE

Organ in queen in which sperms are stored.

SPERMATOZA

Male reproductive cells.

SPIRACLES

External openings of tracheae through which bees breathe.

SPRING DWINDLING

A condition in which the colony population decreases in size during spring at which time exponential population growth is anticipated.

STAMEN

Male part of flower on which pollen-producing anthers are borne.

STING

Modified ovipositor of female Hymenoptera developed into organ of defense.

SUCROSE

Cane sugar; main solid ingredient of nectar before inversion into other sugars.

SUPER

A wooden box with frames containing foundation or drawn comb in which honey is to be produced. Named for its position above the brood nest. The same type of

box is referred to as a hive body when it is situated below the honey supers and is intended to be used for brood rearing and pollen storage.

SUPERSEDURE
The replacement of a weak or old queen in a colony by a daughter queen—a natural occurrence.

SUPERSISTERS
Queens or worker bees produced by a single queen and sired by identical sperm from a single drone (subfamily).

SURPLUS HONEY
A term generally used to indicate an excess amount of honey above that amount needed by the bees to survive the winter. This surplus is usually removed by the beekeeper.

SWARM
Natural division of colony of bees.

T

TARSUS
Fifth segment of bee leg.

THORAX
Middle section of a bee.

TRACHEAE
Breathing tubes of insects.

TRACHEAL MITE
(See Acarapis Woodi.)

TROPHALLAXIS
The mutual exchange of regurgitated liquids between adult social insects or between them and their larvae.

TUMULI
Nest mounds (wild bees).

U

UNCAPPING KNIFE
Knife used to remove honey cell caps so honey can be extracted.

UNITE COMBINE ONE COLONY WITH ANOTHER.
Unsealed brood Brood in egg and larval stages only.

V

VARROA DESTRUCTOR
An external parasitic mite that attacks honey bees Apis cerana and Apis mellifera.

VIRGIN QUEEN
Unmated queen.

VSH (VARROA SENSITIVE HYGIENE)
USDA ARS scientists Dr. John Harbo and Dr. Jeffrey Harris at the Honey Bee Breeding Laboratory in Baton Rouge, Louisiana, have defined and tested a trait of the honeybee which appeared to suppress mite reproduction (SMR). Recently it has been better defined as "varroa sensitive hygiene (VSH)." This is a form of behavior where adult bees remove pupae that have reproductive mites but do not disturb pupae that have mites that produce no progeny.

W

WALK-AWAY SPLIT
Frames with eggs and worker bees are removed from a queenright hive and installed into an empty brood chamber or nuc. The bees should create a queen cell out of a suitable egg. Once the queen hatches, successfully mates and returns to the hive, the hive will be queenright. Another option is to remove one complete brood chamber from a hive that has newly laid eggs in it, including bees, and move to a new location for the start of a new hive.

WASHBOARDING
Worker honey bees exhibit a "group" activity known as rocking or washboarding on the internal and external surfaces of the hive. This behavior is believed to be associated with general cleaning activities but virtually nothing is known as to the age of worker engaged in the behavior, under what circumstances workers washboard and the function of the behavior. Washboarding behavior appears

to be age dependant with bees most likely to washboard between 15–25 days of age. Washboarding increases during the day and peaks through the afternoon. Workers may respond to rough texture and washboard more on those surfaces. The function of this behavior remains to be elucidated.

WAX GLANDS
Glands on underside of bee abdomen from which wax is secreted after bee has been gorged with food.

WAX MOTH
Lepidopterous insect whose larvae destroy wax combs.

WILD BEES
Any insects that provision their nests with pollen, but do not store surplus edible honey.

WINTER CLUSTER
Closely packed colony of bees in winter.

WIRED FOUNDATION
Foundation with strengthening wires embedded in it.

WIRED FRAMES
Frames with wires holding sheets of foundation in place.

WORKER BEE
Sexually undeveloped female bee.

WORKER COMB
Honeycomb with about 25 cells per square inch.

WORKER EGG
Fertilized bee egg.

NOTES

NOTES

www.ingramcontent.com/pod-product-compliance
Lightning Source LLC
Chambersburg PA
CBHW051917210326
41597CB00033B/6170